A LITTLE BOOK OF THOUGHTS:

COMMUNITY DEVELOPMENT

A Little Book of Thoughts:
Community Development

Robert Hughes can be contacted at: simpleconversations13@gmail.com

ISBN: 978-06488978-2-8

First published 2022

A LITTLE BOOK OF THOUGHTS:

COMMUNITY DEVELOPMENT

ROBERT HUGHES

A special thank you to those who have inspired me to write and
print this book:
Jane for the challenge,
Richard for the encouragement, and
Remo for his patience.

PREFACE

The thoughts presented in this book come from a lifetime of involvement in community. Growing up and raising a family in a small Australian rural community provided opportunities to play, coach and perform executive roles in sports clubs and associations, and to manage and maintain a local recreational facility. This provided valuable experience in communicating and collaborating with community members and groups.

The skills and knowledge gained from my earlier farming and community life were strengthened during 15 years international work managing private businesses and developing, delivering, and evaluating Non-Governmental Organisation and UN projects in Africa and the Pacific.

Returning to Australia in 2012 my roles dealt with settlement and community projects in the Northern and Western suburbs of Melbourne. It also included co-designing and delivering a strategy designed to encourage people from culturally and linguistically diverse communities to take part in the 2016 Australian Census.

These experiences have strengthened my belief that the foundation on which successful community development activities are built is strong and robust relationships. Ones which allow people to agree to disagree without damage to the relationship.

I believe these relationships begin during initial contact and are constructed following three simple principles:

1. Treat others with respect;
2. Be open and honest with others, letting them clearly know what you can and cannot do; and
3. Display a desire to understand and learn from, and about others.

Following these principles encourages open sharing of thoughts and ideas and more willing collaboration.

What? Why? How?

When wanting people to embrace ideas, accept changes or take on challenges, explain to them what, why and how. This is a simple mantra for work, community, social and personal situations.

Explain what the idea/change/challenge is that needs to be introduced or met.

Educate/motivate why there is a need to accept, embrace and work towards the idea/change/challenge.

Show how to deliver the idea/change and meet challenges faced, by providing people with relevant knowledge and skills.

Village resource mapping – Vanuatu

Community consultation – Vanuatu

Ground water catchment - Somaliland

Community consultation - Vanuatu

Do not become part of the problem

Challenging the community to learn new skills, take ownership and drive projects themselves, takes time and effort. It is often easier and quicker to do things yourself but, that does not strengthen individuals or communities nor lead to sustainability of activities.

Challenging, mentoring, building capacity and encouraging communities to take control, strengthens individuals and communities while reducing their dependency on external assistance.

Address causes, not symptoms

When looking at issues presented by the community, dig deeper to identify the cause of the issue.

Treating the cause will deal with the problem, treating symptoms allows problems to continue.

You are an outsider

Build strong and trusting relationships, learn from communities and get close to community members. However, keep in mind that you are always an outsider.

It can be insulting to the community and/or risky for yourself to think otherwise.

Documentation is an essential and important program component

Field workers often see practical work with communities as productive whilst considering recording and documentation as basically unproductive. This generally leads to poor or untimely documentation and recording.

It's important not to forget that poor recording often leads to unsuccessful project applications. This ultimately leads to you not having a job, which in turn restricts your opportunities to support communities.

The exit strategy is considered at each stage of a project's planning and delivery

A clear exit strategy helps focus on community ownership and the need for a strong capacity building component in the project. This provides the community with knowledge and skills that allows it to successfully take over activities.

It is also a reminder that projects are not 'ours' and, for them to be successful, external support should be provided within a limited timeframe.

Projects do not establish community groups

Build on existing groups and, if a group is required and does not exist, it is up to the community to want to form it.

Your role is linked to governance training, mentoring and follow-up support. This allows the groups to successfully manage themselves and activities in a transparent and accountable manner.

A community is ultimately responsible for managing its own issues

External support complements what the community already has and is willing to do, it is not the instigator or driving force.

Actions and sustainability are in the hands of the community. Your role is to help prepare them to manage the issues, not to be the solution.

Respect the time of community members

Involvement in community activities is only
one part of a person's life.

Organise meeting and activity times in
collaboration with community members to fit
within their and your availability.

Community ownership is critical

Without the willingness of the community to be involved and take control of activities there is no sustainability.

Give true ownership, not just 'talk' ownership where you give with one hand then remove with the other.

Be brave, allow others to make errors and learn. Take care not to step in and take over because you 'know' what is best and good for others.

Shallow-well hand pump – Somaliland

Island hopping - Vanuatu

Sanitation small business – Vanuatu

Moving between villages - Vanuatu

Identify the skills, knowledge and resources of a community and its members

It is important to focus the first discussions with individuals and communities on what they have and what they can do. Not what they do not have and what they cannot do.

Bring positivity to individuals and communities by facilitating self-identification of their specific resources, knowledge, and skills.

Identify and consider the perspectives of others

There are always different ways to look at the one thing, ALWAYS!

Learn to look at things while considering historical viewpoints and the present situation of the person/community you are dealing with. This will remove both the community's and your biases.

Sustainability comes from building community capacity

Activities stop, groups disband and buildings crumble however, the knowledge and skills gained by community members remains with them and the community after projects end.

Building the skills and knowledge of your staff, and community members, groups and leaders, is an investment in activity sustainability.

Information is best gathered through conversation not questioning

People provide information freely when they are involved in a relaxed discussion, not when face with a formal series of questions.

The skill required to gain the information is to ask questions in a way that the person does not feel like they are being questioned.

The key to this informal questioning is knowing exactly what information is required, and then managing the conversation to gain this information.

Capacity building is not just workshops

Capacity building involves identifying the skills and knowledge people require, then engaging them in training or experiences that will provide those skills and knowledge.

Once the skills and knowledge are gained, it is important to provided opportunities for people to test and practice them while receiving experienced post-training support. This builds confidence in their able to use these new skills and knowledge.

Building relationships takes time

Strong stakeholder relationships are the core of successful community projects. Time and effort are required to build, maintain and strengthen these relationships.

Stakeholders include community members, leaders and groups, local and external organisations, governments, businesses, project teams, project evaluators, your organisation's management and funders.

Actively seek similarities, not differences

When comparing diverse groups of people, it is important to acknowledge that there are more commonalities than differences. An effort is required to look for them.

Commonalities provide an affinity with others and are the foundation of building understanding between people.

The danger of identifying and focusing on differences is that it stimulates negative thoughts and bias.

You have a strong influence over others

Your said and unsaid words, taken and not taken actions and attitude, greatly influence those around you. It is important these words, actions and attitudes inspire positivity and confidence in community members.

Do not underestimate the power of words

Words are powerful and have a strong impact on people. They can motivate, demotivate, provoke, challenge, insult and encourage.

Be aware that some languages may not have an equivalent of an English word or, that some words may not translate with the same meaning.

There is a responsibility to be sure that, when using words, your meaning of what you say is what people actually hear and understand.

Speak in words that
are understood by
those you're speaking
with

Presentations are to provide people with information they can understand and use, not to show how smart you think you are.

Keep the language and words you are using at a level people you are talking to understand. This is not 'dumbing down', it is showing respect to the audience.

Village water source – Vanuatu

Local garage business – Somalia

Town water supply construction - Somalia

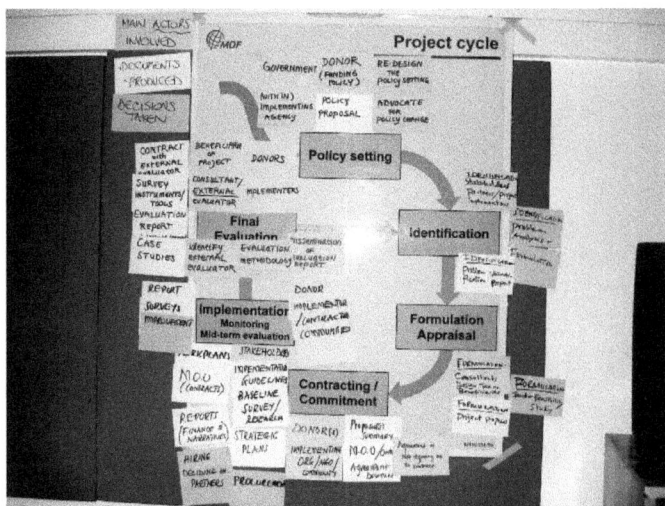

Project cycle training - Somalia

Set work/personal boundaries

You are working with the community, not working for the community. Community members need to be respectful of your time away from work.

If approached by a community member to discuss work at a social function, explain that you are not working. Be firm, set an official meeting appointment for a later date, then move on and enjoy the function.

Strive for continuous improvement

For yourself, your team, and the community, celebrate achievements then move on and strive to do better.

Wherever you can, encourage and support others to grow and develop, this is the key purpose of community development.

Have confidence, yet retain humility

Confidence leads to action, it allows you to attempt things, take calculated risks and be responsible for your actions.

Humility, which includes giving due recognition to the work of others, keeps you mindful of who you are and where you came from. It also keeps the focus on the community, not you.

Making decisions is not difficult

Decision making requires you to first know exactly what you want as the outcome. From there it is a process of gathering information and discussing options with others to reach an informed decision on the best option.

Act on the decision and, if you do not reach the desired outcome, repeat this process until you either reach it or, achieve what has been redefined as the acceptable outcome.

It is better to decide and move forward rather than wait for the 'right' decision before acting.

It is the process of making the decision and acting on it that is often more beneficial than the decision itself.

You're an instrument of incrementalism

It is not possible to make big changes quickly. However, small things you do complement the small things others have done/are doing/will do. Collectively these small changes gradually build into something much bigger.

Awareness of the small role we play in the collective allows an inner calmness that recognises and accepts the slowness of change.

Not achieving what you planned is part of the process of learning, it is not failure

Failure is the decision not to try.

Not succeeding as planned only means that adjustments are required, and another informed decision needs to be made.

Fear of not being successful should not prevent decisions being made or, lead to activities not being attempted.

Understand and acknowledge achievements gained towards reaching overall objectives

Community development work takes time with overall objectives reached through a series of small achievements, one after the other.

Identifying and acknowledging each achievement with the team and community provides motivation and positivity.

Only focusing on reaching the overall objective results in a perceived lack of progress, which can lead to disappointment and despair.

Acknowledge what you do not know, then find out

There is no shame in not knowing. Whereas, pretending to know creates problems.

Community members quickly identify when you are 'faking it' and you not only lose your integrity, but also their trust.

Both are difficult to regain and their loss weakens community relationships and limits sustainable outcomes.

Expect criticism

Humans naturally find it easier to criticise than acknowledge something well done.

Listen to the criticism, analyse it, make adjustments if required and then move on.

Do not be consumed by, or make adjustments for, unjust criticism.

Remain objective

.

Be aware of the potential for decisions to be made and actions taken, that are based on biases, personalities, a desire to be liked, political pressure, friendship or ethnic/cultural pressure.

Be true to yourself and ensure discussions are based on issues, not people.

Enjoy work

To manage work pressure, regularly reflect on the parts of your work that provide you with the most enjoyment and satisfaction.

Look for positives, they are always there, and emphasise these to yourself and the people around you.

Keep in mind that your mood, words and actions have a strong effect on the motivation and wellbeing of others.

Village engagement – Vanuatu

Community village water point- Somalia

Erosion control - Somalia

Farmer workshop - Somalia

Community consultation - Somalia

Field team - Vanuatu

A community is defined as a group of people living in the same place or having similar characteristics, attitudes or interests.

Thus, a broader community that is linked by location comprises a range of smaller communities connected by such things as work, sport, ethnic background, religion, education, age, hobbies and interests. It is important to be aware that at times, this will lead to competing needs and interests of groups and individuals within the groups.

Community development work is not complicated, but it is also not easy. The complexity of dealing with issues that result from the communities within communities dynamics makes community development work a mixture of challenge, frustration, satisfaction, interest and reward.

As you undertake the work you will come across people who are willing to help you. When you become aware of these people, accept what they offer. In turn, be open to sharing your experiences and knowledge with others to support their professional and personal development.